Rainbow Beaches

by Lara Winegar

On this beach in India, waves may crash on the shore. Small rocks can tumble in the waves. Weathering will break the rocks down. Over time, they will become tiny grains of sand.

You might have seen a tan beach.
But have you ever seen a **black beach**?

The rocks near this beach in Iceland are black. Weathering changed the rocks. Waves slammed into them again and again. Waves wore rocks into grains of sand. The minerals in the sand make it black.

Now you've seen a black beach.
But have you ever seen a green beach?

This beach in Hawaii has green sand.
It is made of a mineral called olivine.
The olivine formed in volcanic rocks.
Crashing waves wore down the rocks.
Waves turned the rocks into sand!

Now you've seen a green beach.
But have you ever seen a red beach?

The cliffs and sand on this beach in Australia are red. Ocean waves smashed into the cliffs. The waves broke off bits of rock. These red rocks formed the red beach.

Now you've seen a red beach.
But have you ever seen a pink beach?

Not all sand is made of weathered rock. This pink sand beach in the Bahamas is made of pieces of tiny animals. These tiny animals live in the ocean.

Some of these animals have dark red shells. When they die, their shells are broken down by moving ocean water. Then waves carry bits of their shells to the shore. Over time, the shells become sand.

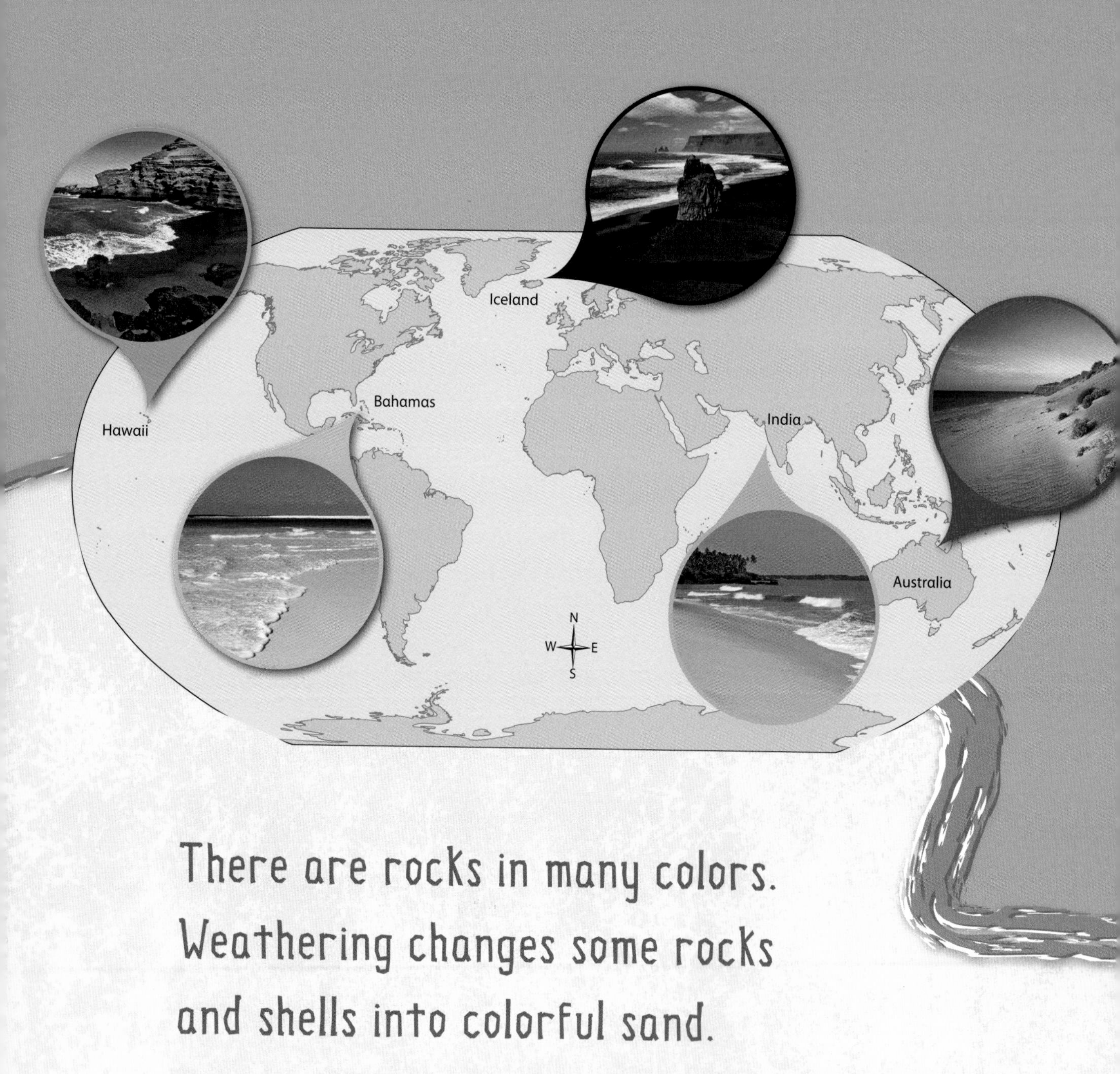

There are rocks in many colors.
Weathering changes some rocks
and shells into colorful sand.
The world is full of rainbow beaches!